Goodnight Moon, Hello Stars

A Gentle Bedtime Science Story

By Rudo Kashiri

Curious Minds Explorers Publishing

Goodnight Moon, Hello Stars

A Gentle Bedtime Science Story

Illustration by Prabir Sarkar

Published by Curious Minds Explorers Publishing

Stories That Launch Curiosity

www.curiousmindsexplorerspublishing.com

ISBN: 979-8-9996203-8-5

Printed in the United States of America

This book is a work of fiction inspired by real science.
Names, characters, places, and events are used to encourage curiosity and learning and are presented in a fictionalized format.

This book belongs to

To Tafadzwa and Rebone,

Parents are the first teachers in their children's lives. You demonstrate exceptional care in ensuring your children are exposed to the very best, and you consistently encourage them to grow into the best versions of themselves.

Thank you for filling your children's nights with laughter, stories, and love, and for creating a nurturing foundation where curiosity, learning, and wonder can flourish.

"I'm always interested in learning something new."
— Katherine Johnson

Meet Itai, Jack, and Ambuya.

It's nighttime. The sky is full of stars.

"Let's look at the Moon," says Ambuya.

The stars wink at us from near and far.

It's nighttime now. The stars shine bright,

Itai brushes his teeth in soft, calm light.

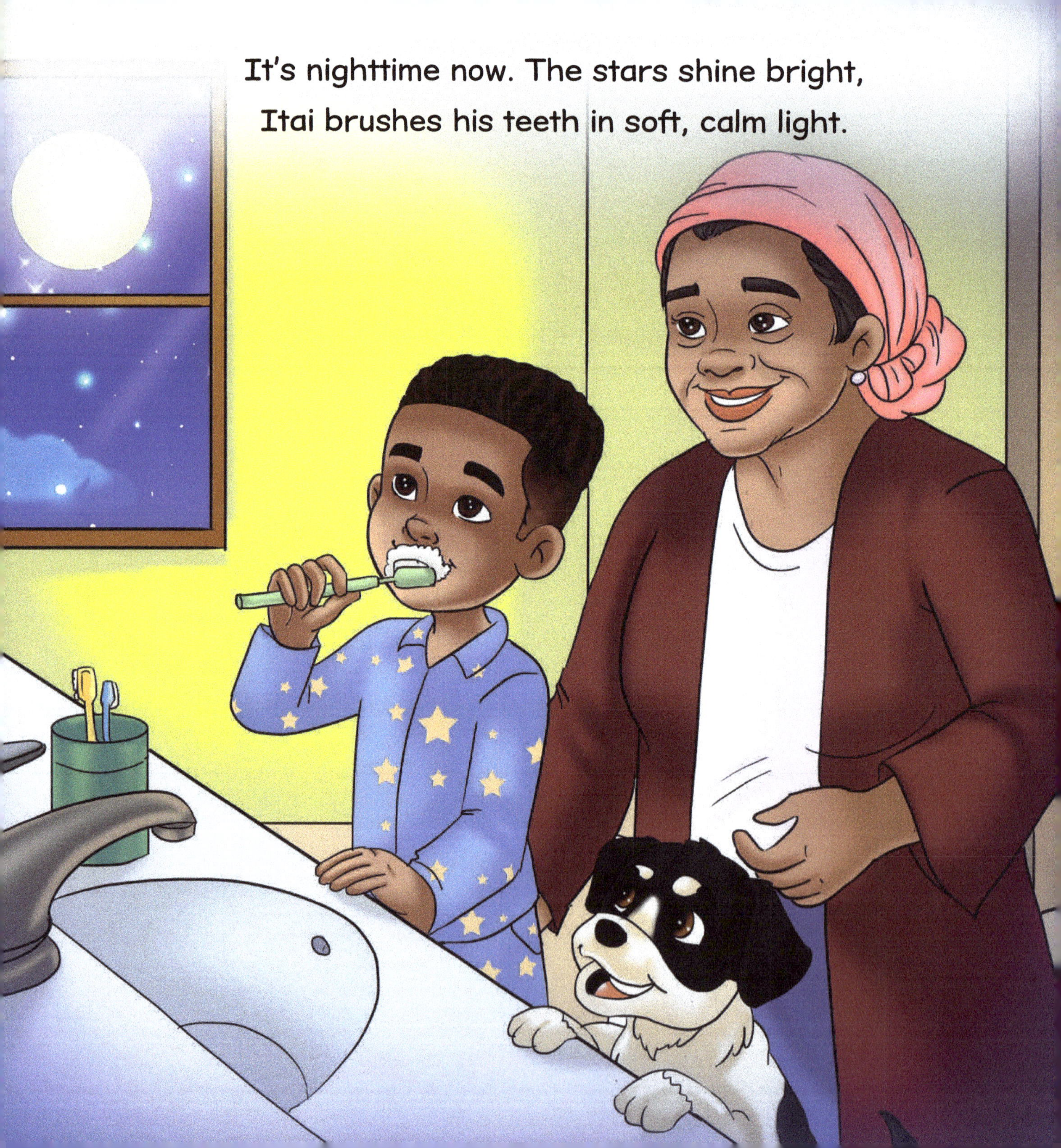

"Ambuya, why is it dark?"

"The Sun is shining on the other side of the Earth right now," says Ambuya.

"Will it wake up?"

"Yes," says Ambuya.
"The Sun comes back in the morning."

The Moon shines bright up in the sky.

"Goodnight, Moon!" says Itai.

"Wow! The Moon is so big!"

"Goodnight, Moon. I love you."

"Can I touch the Moon?"

"The Moon is far, far away," says Ambuya.

"Maybe I can fly to the Moon one day!"

Itai looks up high. "Look! Stars in the sky!"

The stars twinkle, twinkle. Bye-bye!

"Hello, stars! Why do you twinkle?"

"Can I count the stars, Ambuya?"

"There are so many stars!" says Ambuya.

"Let's count: one, two, three...
there are too many to count!"

"Stars are far away.
They sparkle in the night sky."

Ambuya smiles and shows the Moon.
Its shape changes, round and
then thin, like a smile.

Can you make a round Moon with your hands?

"Why does the Moon change shapes?"

"Sometimes the Moon looks big and round," says Ambuya.

"Sometimes we see just a little part. But the Moon is always there."

"I like the big, round Moon best!" says Itai.

Ambuya holds Itai tight.

Sleepy eyes, cozy night.

"Goodnight, Itai. Sweet dreams tonight."

"Goodnight, Ambuya. I love you."

"Goodnight, sky. Hello, sleep tonight."

Itai snuggles in bed.
Mom and Dad read a book.
Ambuya smiles nearby.

Do you have a favorite bedtime story?

"We love you, Itai. Sleep tight," says Mom.

"Goodnight, Itai," says Dad.

Jack circles once, then settles at Itai's feet.

Itai closes his eyes, so sweet,

Jack curls up beside his feet.

Moonlight fills the room.

Stars say, "Goodnight."

Itai starts to dream.

The Sun comes up, so warm and bright.

Itai wakes up. Good morning, light.

"Good morning, Jack! The Sun is back!"

"Let's go play!"

Jack stretches by his side.

"Hello, Sun. Good morning, day!"

After breakfast, out they go.
Itai throws the ball.
"Run, Jack! *Go, go!*"

Do you like to play outside?

"Catch the ball, Jack!"

They laugh and play in morning light.

"I like the Sun. It makes me smile," says Itai.

"Hello, sunshine. Hello, new day."

Author's Note

Goodnight Moon, Hello Stars was inspired by the calm, comforting moments of bedtime and the wonder children feel when they look up at the Moon and stars.

This story celebrates curiosity, imagination, and the special time shared between children and the people who tuck them in each night.

As you turn the page and whisper goodnight, I hope you'll pause to look up together and wonder:

- Where is the Moon tonight?
- Which stars can you spot?

May this story be a gentle reminder that every night sky holds new questions and that love is the coziest place to explore them.

Rudo Kashiri

Curious Minds Explorers Publishing

www.ingramcontent.com/pod-product-compliance
Lightning Source LLC
LaVergne TN
LVHW070900210926
848482LV00016B/985
9798999620385